FIND
THE ANIMAL FOOT PRINT
FIND THE footPRINT
THAT
GOES WITH each of these ANIMALs

ANSWERS IN BACK
OF BOOK

FEATS of FEET

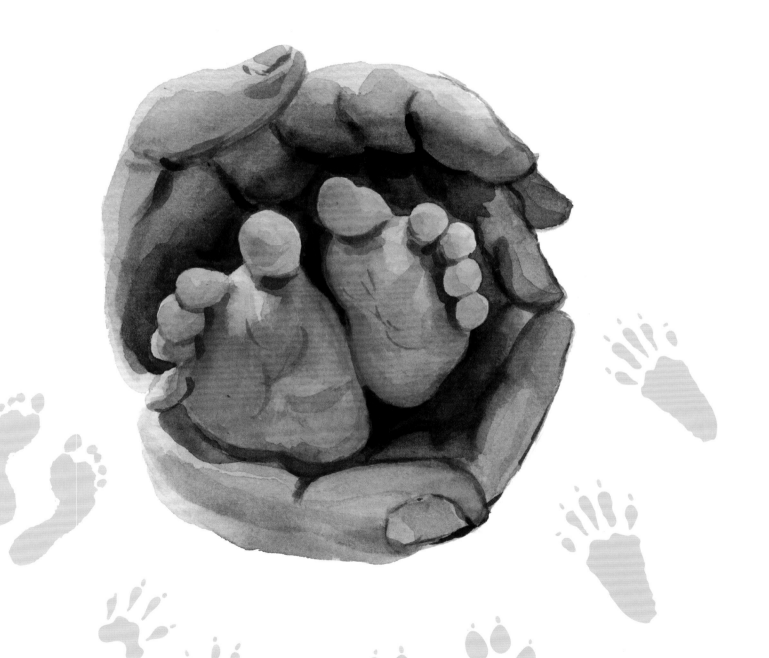

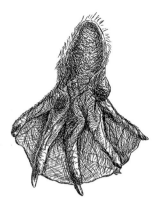

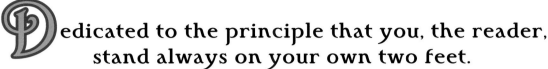 edicated to the principle that you, the reader,
stand always on your own two feet.

·Acknowledgments·

Zenaida Vega for rounds and rounds of typing of the manuscript,
Maureen Slater for secretarial skills,
Zachary Peare for serving as a pollster of his second grade class
to determine which of several possible covers was the most appealing,
and Betsy Stevens for coordination of all parties engaged in this labor of love.

For information write:

Ardor Scribendi, Ltd.

145 East 32nd Street, 10th Floor

New York, NY 10016

Tel: 212.448.0083

Visit our bookstore at www.ardorscribendi.com

ISBN 1-893357-40-6

Printed in the United States of America

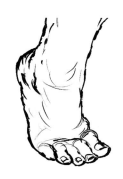

OREWORD

"Man's foot is all his own. It is unlike any other foot. It is the most distinctly human part of his whole anatomical make up. It is a human specialization and, whether he be proud of it or not, it is his hallmark and so long as Man has been Man and so long as he remains Man, it is by his feet that he will be known from all other members of the animal kingdom."

Wood Jones
(Hall, Michael C. Locomotor System. Functional Anatomy.
C Thomas Springfield, 1965:437)

This book was inspired by a trip that the authors, colleagues in medicine, took to Denali National Park in Alaska in June 1994. Their hope is that by reading and hearing a story like this one, young people will be inspired to learn not only to recognize the importance of each of the parts of their body, but also to be motivated to care for and about them.

COME WITH US ON AN EXCITING JOURNEY TO THE WORLD OF FEET

We can even walk to get there! You use your feet so much that you don't even think about them unless you stub your toe! But have you ever thought about how unique your feet are and how much you use them? All animals and birds have feet that serve their needs just as do yours, but their feet differ from yours in the way they look and in many other intriguing ways.

The function of feet, that is, what they do for you and for other animals, is revealed by the form of the way they are made. Feet can do many things because they are designed in different ways.

These are some of the things your feet help you to do:

Give support by holding up your entire body, whether you are tall and thin or short and round.

enable locomotion by allowing you to move in many ways, such as walking, running, jumping, climbing, and swimming.

· 6 ·

aintain balance by keeping you from falling down because both of your feet can be planted firmly. You can even stand on one foot - if only for a short time. But two feet are better! Some animals need more than two feet to maintain balance.

Provide a sense for touch by letting you feel the difference between a hard wooden floor, a fuzzy carpet, and soft sand on the beach. If you have ever had a splinter in a foot you know how sensitive to pain feet can be and how effective they are at sounding an alarm to objects that can injure you. If you now look carefully at how feet are made, you will understand better how they function.

Feet of different creatures that live on land, in the sea, and in the air, are made in many different ways. We human beings are the only animals whose feet are arched, that is, the entire bottom of them (it's called the sole) does not touch the ground because the inner part of it is raised like the arch of a building. We also are the only ones whose first toe is longer than the other toes.

Spread your toes and see how by doing that you gain even more support. Look and see where your big toe is. Is it on the inner side or the outer side of your foot? It's on the inner side because being there makes it easier for you to walk and keep your balance.

Look at the footprint of a bear. Is its big toe on the same side as yours? No! It's on the outer side! A bear has such a big body it has to walk in a special way in order to maintain its balance. Having a big toe on the outer side of the foot helps a bear to do that. Your feet are made just right for you, like holding you up when you are standing. The hard bones and the softer parts of your feet help you do this.

The arch permits your feet to spring up and down when you walk. The soft padded covering of skin and fat immediately below it protects your bones from hurting when you walk, run, and jump. When you stand, you don't even think about your feet, even though all of your weight is on them!

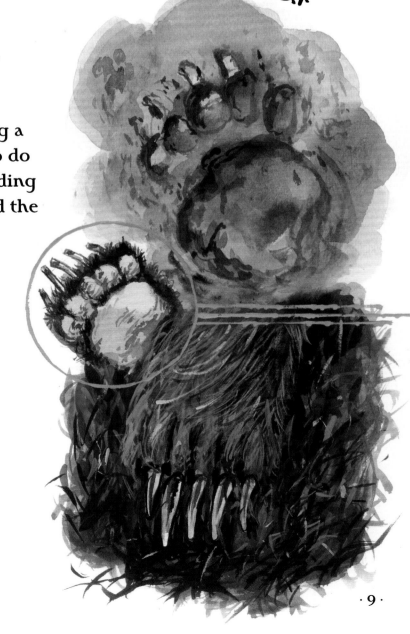

Bear

What a humongous job an elephant's feet have to do! The elephant is the heaviest of all animals that live on land, and the whale is the heaviest of animals that live in the sea. But whales do not have feet!

Because elephants are so heavy, they have huge pads under their toes that cushion their feet and keep them comfortable, just as sneakers on your feet do for you. Do you know that elephants actually walk on their toes? They do! You know what else?

The large pad on the bottom of the toes helps an elephant to walk quietly—just like you sometimes do when you take off your shoes and tiptoe, or when you want to sneak up on one of your friends to surprise them.

Bottom of Elephant Foot

· 10 ·

Have you noticed how hard it is for you to walk on wet sand on a beach, on soft fresh snow, or on soggy mud? Have you wondered why that is? Or have you ever said to yourself, "If I only had bigger, wider feet they would help me from sinking into snow or mud"? That is why thousands of years ago someone invented snowshoes.

Can you think of any animals or birds that have their own built-in snowshoes which keep their feet from sinking into snow or mud? Have you ever heard of snowshoe rabbits? They have big back feet with thick fur (which is the name for hair in some animals) on the pads to prevent them from sinking into snow or mud. Their back feet are shaped like snowshoes.

Snowshoe Rabbit

Look at the feet of the State bird of Alaska, the Ptarmigan. Believe it or not, in the winter they actually grow feathers on their feet and toes!

Ptarmigan

Those feathers help keep them on top of the snow, just like snowshoes do for you.

Moose have feet with large split hooves that keep them from sinking when they walk on mushy ground and mud.

Moose

Different animals have different kinds of feet and the differences in each are just right for them because they help them to do what they do best.

Camel

How are camels able to walk so easily over the sandy desert? It is because of their feet! Their feet are very different from yours. They have only two toes, the bones of which are wide apart.

Those bones sit on a broad cushion-like pad that spreads out as the camel walks and keeps that hump-back animal from sinking into the soft sand. Riding camels is the favorite way we humans travel in the desert.

Besides walking, running, and jumping, what other fun things can you do because you have feet? Can you name some? What about climbing and swimming?

Your feet are designed in such a way that they let you walk upright, which means standing up straight when you walk. Try taking a few steps.

Did you notice that with each step the bottom of your heel (that's the back part of your foot) touches the floor first? Very soon after that the front part of the bottom of your foot (that's the "ball" of it, just behind your toes) hits the floor next.

Mountain Lion

Mountain Lion

Once you learn how, it's pretty easy to walk (you don't even have to think about it) and that is because your feet are shaped so well. Look at the bottom of one of your feet.

Do you see that the heel is wide and has a thick pad that can touch the floor gently? Look just behind your toes and there's the ball of the foot. It is even broader and better padded than your heel. It needs to be like that so with every step you take it can support the weight of your whole body. Your foot, just like that of most other animals, is designed especially for walking. Mountain lions and other big cats have rubbery pads on the bottom of their feet, those pads being a much bigger version of the pads of skin and fat you have on the bottom of your feet. Those pads enable big cats to walk comfortably and, most important for them, quietly, so that they can surprise the unsuspecting creature they will eat for breakfast, lunch, or dinner.

15

By the way, if brand new shoes are too tight, they can cause blisters to develop on your feet.

Doctors call those "friction blisters" and if shoes do not fit properly and are worn for a long time, parts of the feet can become thick and rough (doctors call those calluses).

Big cats run much faster than you can because in order to eat they have to outrun and catch other animals, like antelopes, who run very fast.

Tiger

And because of the big pads underneath their toes, big cats don't have sore feet even when they run very fast. You know that when you run with bare feet on a hard floor or sidewalk your feet sometimes hurt, especially if you run fast and for a long time.

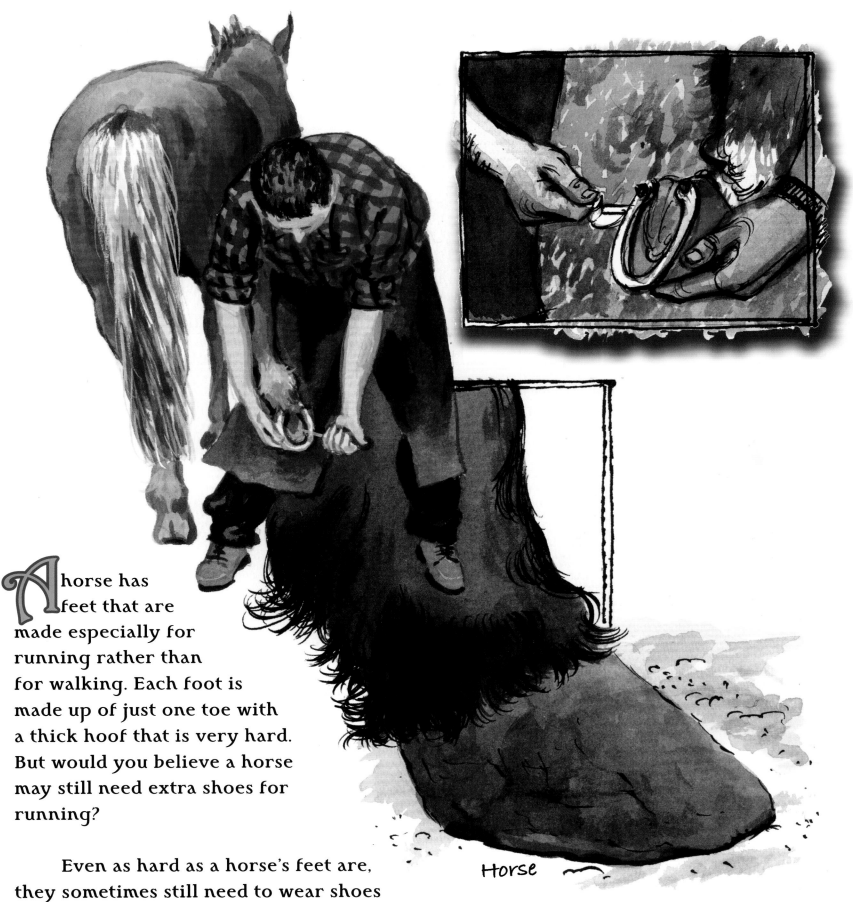

Horse

A horse has feet that are made especially for running rather than for walking. Each foot is made up of just one toe with a thick hoof that is very hard. But would you believe a horse may still need extra shoes for running?

Even as hard as a horse's feet are, they sometimes still need to wear shoes made of metal. Those "horse shoes" keep their hard hooves from cracking, especially when they are galloping over rocky ground.

If you wear "running shoes" that have thick pads on the bottom of them, your feet, like those of big cats, won't hurt at all.

How about the feet of an ostrich? The ostrich is the only bird in the world with only 2 toes, yet it is the speed champion of all birds that cannot fly. But can they ever run fast! On the bottom of their big toe they have prominent treads, like those on the tires of a car, which helps propel them as they run and keeps them from slipping.

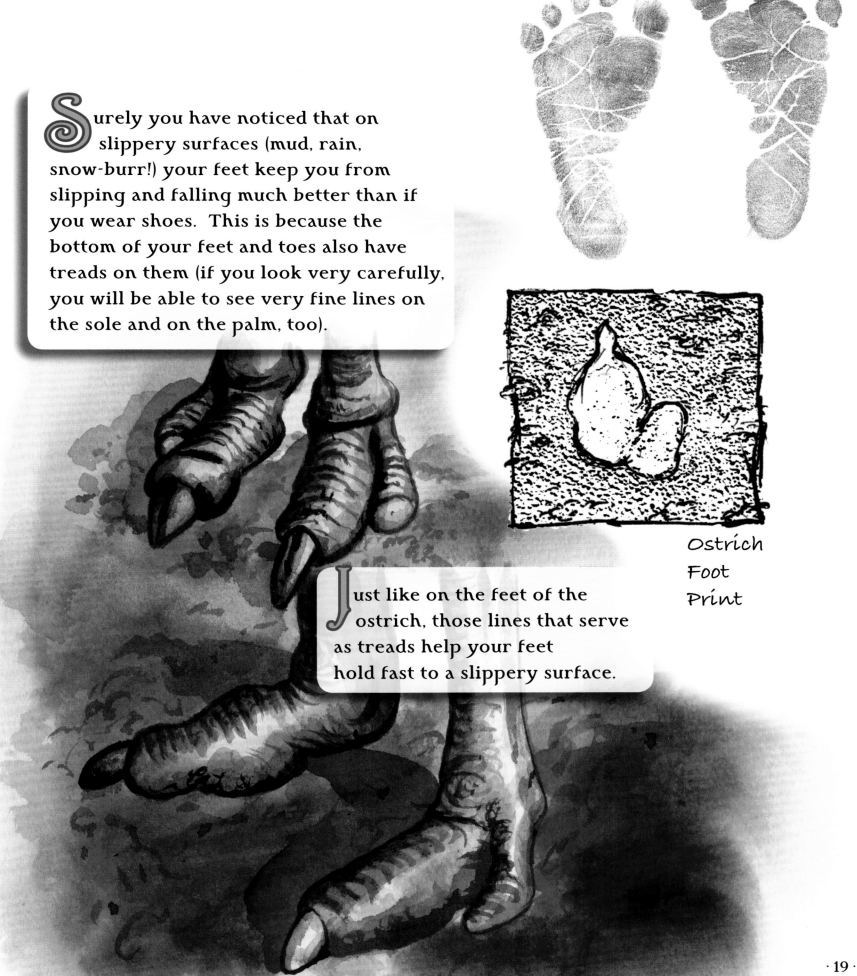

Surely you have noticed that on slippery surfaces (mud, rain, snow-burr!) your feet keep you from slipping and falling much better than if you wear shoes. This is because the bottom of your feet and toes also have treads on them (if you look very carefully, you will be able to see very fine lines on the sole and on the palm, too).

Ostrich
Foot
Print

Just like on the feet of the ostrich, those lines that serve as treads help your feet hold fast to a slippery surface.

· 19 ·

Have you ever tried going down a slide with your bare feet under you? What happens? You either won't move at all or you will go very slowly. Your toes spread out and your feet grip the board. Now if you add a little water to the slide — away you will go!

Many of the shoes you wear are made with treads on the bottom of them that help you grip a slippery surface better than you can with your bare feet alone. Some shoes are even made for special places, like sneakers that grip the hard wooden floor of a basketball court, or hiking boots that grip rock and snow, or climbing boots that grip the side of a mountain.

Animals don't slip because their feet prevent that from happening. Let's look at the feet of some animals and see how they work. Polar bears, for example, can walk and run on ice without losing their hold. The reason is that the bottom of their feet are furry. Besides keeping their feet warm (like shoes do for you), the fur grips the ice and keeps them from slipping. Those furry feet also help them to walk quietly so they can surprise a seal resting on the ice before it can escape by diving into the water and swiming away.

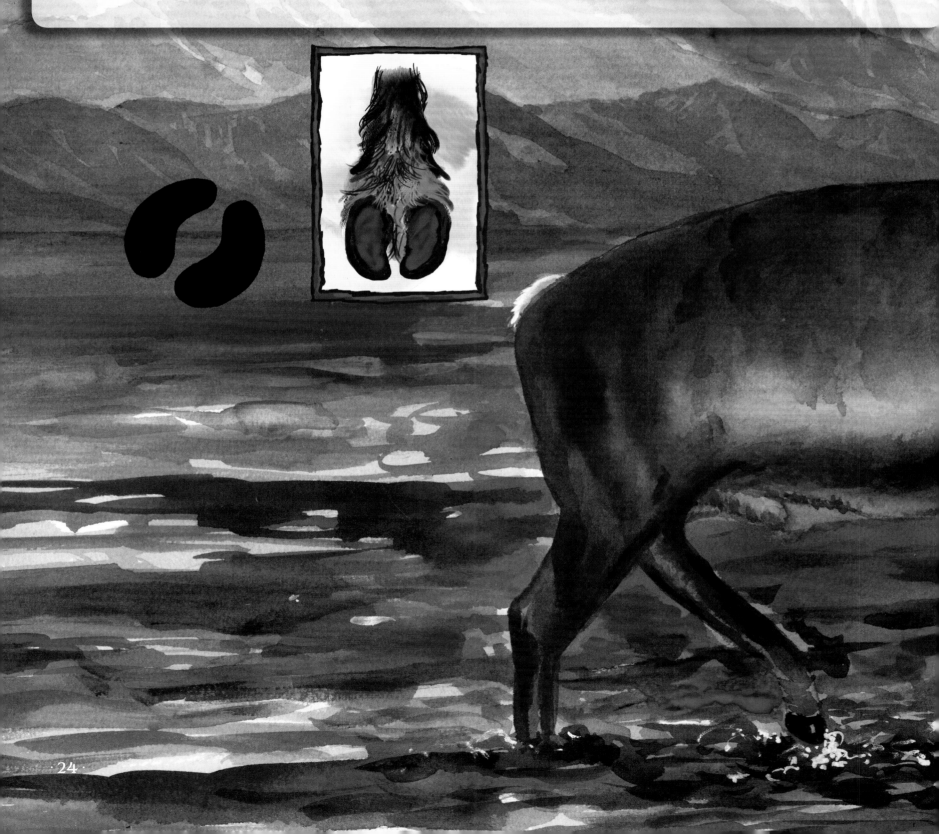

Caribou, who look like Santa Claus' reindeer, also have special feet. They live in the Far North where lots of snow is on the ground most of the year. Each foot of a caribou has 4 toes, 2 large ones and 2 smaller ones. The large toes spread apart on soft snow and that keeps their feet from sinking in. The smaller toes help even more than the large ones by catching hold if the larger ones start to sink in. The outer edges of the hooves are sharp and hard. Those edges grow longer in the winter to better grip the snow and ice and to dig better through snow in order to find food. When the soft pads on the inside of the horny hooves shrink, the hooves function like suction cups. Hair partially covers the pads to protect them from sharp ice.

The feet of mountain goats are split into 2 toes. The toes spread out and hold firmly to jagged rocks. Pads on the bottom of the toes are hard and rough, but flexible, and they prevent mountain goats from slipping. The outside edges of the hooves also are hard and the tips of the hooves grow beyond the pads.

The edges and tips catch onto bumps and cracks in rock and this keeps them from slipping, even on the side of a very steep mountain. Your bare feet do the same for you when you squat on top of them to keep from sliding down a steep hill covered with wet grass.

If you want to swim like a duck and move through the water much faster than a duck, you can change the shape and size of your feet to resemble those of a duck. Adding flippers makes your own feet webbed like those of ducks.

Animals and birds that spend much of their life in the ocean have webbed hind feet that look like your feet when they have flippers on. That webbing is what makes webbed creatures such great swimmers.

ometimes you use your feet to do things that they are not exactly made for and then they can help you out in a pinch. Have you ever used your foot as a shovel? Yes, a shovel! Have you ever used your feet (especially bare ones) to help your hands pile up sand for a sandcastle at the seashore? And have you ever used your feet to help you rake or pile up leaves in the fall? How about using your feet to shovel snow off the front steps of your home? Have you ever done that? I bet you have. You are able to do that because of the way your feet are made. You can even change your feet in such a way that they act as an even bigger shovel.

A mole has thick claws that can function as a shovel. The feet seem to grow right out of its shoulders. In fact, moles use their feet like two shovels that burrow through the ground, shoveling as they go.

$\mathcal{Y}$ou can teach your feet
to do some very unusual things.
For example, some people actually
write and feed themselves with their feet!
That can be done because of the marvelous way
your feet are constructed.

People who use their feet in this way were either
born without useful hands or their hands were
injured in some way. This is called being disabled.
Those people are to be greatly admired for their
courage and their ability to learn to use other parts
of their body to do the job of parts they don't have
or cannot use. They have learned to write and do
other amazing things with their feet that you
would not at first think possible.

Some animals and birds have feet designed in such a way that they are able to perform some very remarkable tasks. Pandas do not have hands, but instead have special front feet that help them crack tough bamboo stems, which are their favorite food. Their front paw contains a prominent wrist bone that serves like a thumb for gripping the bamboo stem as they bite it with their teeth.

Many birds have very long nails, called talons, that they close tight to catch their food, which is mostly fish. Bald eagles not only have talons, but also "bumpy lumpy" toes. They have bumps on each of their toes that help them hold onto small smooth fish which otherwise would slip through their talons, just like tadpoles slip through your fingers!

Have you ever tried to catch minnows or tadpoles in a pond with your bare hands? If you have, you know how difficult it is to catch even a single one of those slippery creatures!

Penguins use their feet to incubate their eggs! Do you remember where penguins live? Guess! Antarctica. Penguins live on top of snow and ice, as well as in water.

It is so cold in Antarctica that penguins often stand on their heels with their toes off the ice in order to keep those tootsies warm, just as you sometimes do when you have to stand on a cold floor before putting your socks on!

In Antarctica there are no twigs or leaves for penguins to make a nest. They lay their eggs on the ice of the sea. In order for the eggs to hatch, they must be kept warm. So what do penguins do? They keep their eggs warm by resting them on top of their feet! When the egg hatches, the baby penguin warms her feet by standing right on top of her daddy's feet! The feet of daddy penguins have a rich supply of warm blood that keeps them warm.

Birds called "blue-footed booby" live where it is hot and dry all the time. They, too, use their feet to incubate an egg and they do that by placing the egg underneath their feet! Those feet have such a big supply of blood that they keep the egg warm. When the egg begins to hatch, boobies then rest the egg on top of their feet. Until about one month of age, the newborn chick sits on a parent's feet to keep warm. The feet of blue-footed boobies are unusual in another way. They are dazzling blue!

A male blue-footed booby uses his bright blue feet to attract a female so that, together, they can make little boobies. Other animals also use their feet to call attention to themselves. Women human beings attract men by painting their toenails shades of red—and sometimes even blue!

· 33 ·

Otter

Speaking of color of feet, have you ever noticed how your own feet change color? What happens when you spend time in the sun? The skin darkens to various shades of brown (it's called a suntan). Some feet are naturally brown.

White-Tailed Deer

Cells in the surface of the skin, known as melanocytes, make a brown pigment, melanin, which is responsible for the color of the skin. Melanocytes in pale skin make little melanin, whereas melanocytes in dark skin make much melanin.

Our trip by foot has come to an end. We have traveled far and wide and hope your feet are not too sore! But even more, we hope you now realize how interesting and useful your feet are to you. Take good care of them so that they can continue to do the many neat things you ask them to do. And remember, too, that feet of different animals were each created in a special way so they could do things they are needed to do. But aren't you glad you have the feet you do and not ones with hairy bottoms, flippers, or talons?

"If I were to make a study of the tracks of animals and represent them by plates, I should conclude with the tracks of man."

Henry David Thoreau

Bob Cat

Muskrat

Glossary of "Feet"
Words and Phrases

Answer with one's feet: to walk out, leave, resign

Cold feet: being afraid to act, at the last minute, in a way expected

Crow's feet: deep wrinkles at the side of the eyes that fan out in the direction of the ears and that vaguely resemble the feet of a crow

Feet first: to be ejecticed, that is, to leave a place not by walking out

Feet of clay: to be weak

Feet on the ground: to be practical, not too dreamy

Fleet afoot: the ability to run fast

Fleet of foot: a fast runner

Foot brake: a brake that is pushed with a foot, like one in a car

Foot doctor: a specialist in the diagnosis and treatment of diseases of the feet; a podiatrist (chiropodist)

Foot drop: a result of damage to nerves in which a foot hangs limp and cannot work at all

Foot fault: in tennis, stepping on the base line while serving a ball with consequent loss of a point

Foot in mouth: to make an embarrassing mistake in speech

Foot loose: free from care; usually used in the expression "foot loose and fancy free"

Foot pace: normal walking speed

Foot pad: a thin bandage that adheres to a part of a foot in order to cover a wound or protect a site from injury

Foot pedal: a gadget that enables a machine to be operated by a foot

Foot race: a race between runners on foot

Foot rope: a rope that is part of a sail for a boat

Foot rot: a sickness that causes rotting of plants

Foot soldier: a soldier in the infantry; one who moves and fights on foot

Foot sore: tired from much walking

Foot stone: a stone put at the foot of a grave

Foot the bill: to pay for something in its entirety

Foot way: a footpath, that is, a path along which one walks

Footage: the length of movie film

Foot-and-mouth disease: a sickness of cattle and deer caused by a virus that affects the mouth and hooves

Football: an inflated ball traditionally made of leather

Footboard: the board at the end of a bed

Footbridge: a narrow bridge for pedestrians

Foot-candle: a way to measure the power of a light

Footed: having a foot or feet

Beaver

Porcupine

Snowshoe Hare

Moose

Footer: foundation of a building

Footfall: sound of a footstep

Foothill: a low hill near the base of a mountain

Foothold: a secure place to put a foot down, as in climbing

Footing: to be secure on your feet

Footle: to act or talk foolishly

Footlights: a row of lights along the front of a stage

Footling: silly and unimportant

Footlocker: a small trunk

Footman: a man who ran on foot beside his master's horse or carriage

Footmark: footprint

Footnote: a brief explanation written in a text, something usually placed at the bottom of a page

Footpad: a criminal who robs pedestrians

Footpath: a narrow path for use by pedestrians only

Footprint: the track left by a foot

Footrest (footstool): a stool to rest the feet on

Footsie: a flirting game or too cozy a relationship

Footstep: a person's step

Footwear (foot gear): shoes, boots

Footwork: a way of moving feet as in dancing

Goosefoot: plants like spinach and beets

Hot foot: to hurry or to light a match that has been maliciously placed between the sole and the leather of someone's shoe

Land on one's feet: to come through a difficult situation safely

Neats foot oil: a light-yellow oil made by boiling the feet and shin bones of cattle

Pussyfoot: to walk quietly or to act timidly

Sit at the feet of: to be an admiring student of

Something is afoot: something not quite right or troubling is about to happen

Stand on one's own two feet: to be independent

Sweep off one's feet: to be very impressed by

Tenderfoot: a newcomer

To get one's feet wet: to become adjusted to a new situation

Two left feet: clumsy

Underfoot: to be in the way constantly

Waiting on hand and foot: helping someone continuously

Human

Dinosaur

Human footprints, that are dating back 20,000 years to the Ice Age have been discovered in a dry lake bed in Australia